THE TWEED TO THE NORTHERN ISLES

Mike Smylie

AMBERLEY PUBLISHING

Acknowledgements

Photographic books such as these are as much about the visual as the writing and therefore an author assumes the greatest gratitude to those who supplied the quality photos. Although some come from my own collection, many do not and therefore I am especially indebted to Michael Craine, Edward Valentine, Ian Murray, Billy Stevenson, David Linkie, Angus Martin, Campbell McCutcheon and The Scottish Fisheries Museum and the other individual members of the 40+ Fishing Boat Association whose names I cannot remember, for their contributions.

General Note to Each Volume in this Series
Over the course of six volumes, this series will culminate in a complete picture of the fishing industry of Britain and Ireland and how it has changed over a period of a hundred and fifty years or so, this timeframe being constrained by the early existence of photographic evidence. Although documented evidence of fishing around the coasts of these islands stretches well back into history, other than a brief overview, it is beyond the scope of these books. Furthermore the coverage of much of today's high-tech fishing is kept to a minimum. Nevertheless, I do hope that each individual volume gives an overall picture of the fishing industry of that part of the coast.

For Ana and Otis

First published 2013

Amberley Publishing
The Hill, Stroud
Gloucestershire, GL5 4EP

www.amberley-books.com

Copyright © Mike Smylie, 2013

The right of Mike Smylie to be identified as the Author of this work has been asserted in accordance with the Copyrights, Designs and Patents Act 1988.

ISBN 978 1 4456 1447 2
E-Book ISBN 978 1 4456 1462 5

British Library Cataloguing in Publication Data.
A catalogue record for this book is available from the British Library.

Typeset in 9.5pt on 12pt Celeste.
Typesetting by Amberley Publishing.
Printed in the UK.

Introduction

The east coast of Scotland has, over the centuries, been responsible for landing more fish than any other part of the British Isles and Ireland. In the late part of the twentieth century, over 60 per cent of fish landed into the UK came from this coast. Charles MacLean, in his 1985 book *The Fringe of Gold*, identifies some 135 fishing villages that dot the mainland coast, with another sixty-nine covering the northern isles of Orkney and Shetland. No other part of the British or Irish coasts boats so much fishing activity. Only Cornwall can come anywhere near in second place.

The title of MacLean's book comes from a quote of James VI (1567–1625), who described his kingdom as 'A beggar's mantle fringed with gold'. Although the king was only referring to the coast of Fife – even then ringed with a series of small maritime villages which fished and traded with Northern Europe – MacLean considered that the description applied to the whole coast between the River Tweed and Duncansby Head.

Today in 2013, Scotland still accounts for over 60 per cent of fish landed, although no statistics account purely for the east coast. However, Peterhead, Lerwick and Fraserburgh respectively remain the three UK ports with the largest landings. Much of the pelagic fish is landed into Lerwick while ports such as Lochinver and Kinlochbervie on the north-west coast have, until very recently, been important ports in respect of landings. Furthermore, other west coast harbours such as Mallaig, Campbeltown, Tarbert and those of the east of the Clyde have traditionally had major fishing fleets.

The 2011 figures suggest Scotland landing some 260,300 tonnes of fish out of a UK total of 397,000 tonnes (65 per cent). Compare Scotland then to England with less than half the landings of 98,700 tonnes (25 per cent of UK total), Northern Ireland 21,900 tonnes (6 per cent) and Wales (approximately 4 per cent). These figures relate purely the landings of UK-registered vessels into the UK. Much is landed into other European ports and, as an example, some 600,000 tonnes of fish in total was landed by UK boats both into the UK and into ports outside the UK, which suggests some 200,000 tonnes was landed outside the UK. This is largely accounted by the fact that nearly half the landings of pelagic fish (herring, mackerel, etc.) by UK vessels were made abroad.

However, go back to 1886 and we see very different figures. That year Scotland received a total of 207,150 tonnes of fish from British vessels, of which almost 75 per cent was herring. England and Wales saw 237,961 tonnes into their ports, of which about 42 per cent was herring. On top of that, another 13 per cent consisted of other pelagic fish such as pilchards and mackerel. However, although England and Wales exceeded Scotland by almost 31,000 tonnes, it has to be pointed out that this English and Welsh tally came from ports as far apart as Fleetwood, Milford Haven, Cornwall, Brixham, London and East Anglia. Scottish landings were still, in the majority although by no means exclusively, from east coast and Shetland ports. However, numbers of fishing boats, from some 10,000 in the late nineteenth century, have declined dramatically to under 1,000, much being due to the European decommissioning policy which has paid owners to scrap their vessels and surrender their licences.

The above figures show that herring was indeed 'king' along this coast. However, before about 1790 the Scots hardly bothered with this rich resource lying close to the coast.

In those days, the Dutch commanded the industry as Scotland had no boats equipped with drift-nets and no market in which to export the fish. They were busy simply catching cod and haddock using long-lines and survived on a pattern of summer fishing and winter farming, as they had done for generations. Any surplus fish was sold locally and any taken out of the locality was cured by drying, salting or smoking.

However, if we look back, we see that there were attempts at establishing a fishing industry. It was the Church that first encouraged full-time fishing when the monks bought the catch in several towns to distribute to the poor. When James II followed suit by offering incentives to catch fish, we can see the beginnings of the building of fishing communities along the coast which today leaves us with this 'fringe of gold' almost along the entire coast under question.

But in 2011 something else changed in the fishing industry. It wasn't just that fewer boats were catching fewer fish, nor that the number of fishermen had declined by 25 per cent to 5,000, nor that cod and haddock supplies were considered under dire threat. Herring quotas had been on the increase, though much of this was going for processing. In Scotland in 2011, for the first time ever, landings of shellfish exceeded those of demersal fish (33 per cent and 30 per cent) with pelagic landings at 37 per cent. Prawns – or more correctly *Nephrops norvegicus*, known variously as the Norway lobster, Dublin Bay prawn or *langoustine* – have become one of the most important species in Scotland.

The fishing industry as a whole can be regarded in terms of four distinct groups: the method of fishing or the Fishing Ways; the Fishing Boats used; those whose occupations are depended on fish – the Fisher Folk; and finally the homeport communities dotted around the Fishing Coast. All these are equally significant in how the industry operates.

Fishing Ways

Given that Scotland as we know the country today has been settled for over 7,000 years, it would be surprising if the coast had not always been a source of food for those living by rivers or the sea. From the archaeological evidence gained from middens, we know that fish and seafood was a major part of the diet of prehistoric man and it is presumed that the shore was the source of most of this food. Shellfish such as mussels, cockles, whelks, winkles, oysters, scallops and limpets could easily be picked as the tide ebbed. Fish could be speared in shallow water and even, as with flounder, kicked from the riverbed. Storms often left fish stranded in small pools or sometimes dead on the sand. Those found in rock pools probably led early man into one of the earliest forms of active fishing in the form of fish traps or weirs – known in Scotland as *yairs* (Scots), *cairidh* (Gaelic) and *cruives* (in the case of salmon weirs). These fish weirs consisted of low stone walls extended up with a wattle structure of interwoven hazel branches between oak posts driven into the stonework and were the forerunners of the stake nets set even today on the beaches. However, these forms of weirs appear more abundant on the west coast. *Cruives* consist of wooden structures built into rivers to catch salmon as they swim upriver and examples can be seen on several east coast rivers.

Fish traps in the form of creels – or pots – laid on the seabed to catch crabs and lobsters are not much more than an extension of the idea of fish weirs. Their use has not changed much in a thousand years.

The use of the line and hook is probably almost as old the fish weirs. Various plants were used to produce line such as nettles and flax, then hemp and cotton. Hooks were made out of various thorns from plants before the advent of metal barbs. These became widely used to catch fish close to the shoreline.

According to Angus Martin in *Fishing and Whaling*, saithe was a most common fish, otherwise known as coalfish or coley. Because of its abundance it was relied upon as a source of food as no other fish was. Not only was there plenty of it, but it was dependable in that it was always available. It was caught by rod and line and also by what was referred to as a *pock-net*, which equates simply to a bag net suspended around a metal hoop which hung down from a pole which in turn, when baited, was immersed in the sea.

It was the arrival of the net that marked a distinct change in the fortunes of fishermen. For a small bulk of material, a strong barrier of a large size could be utilised. They could be easily transported and arranged in varying manners. They could also be set across bays to catch fish on the ebb and then lowered to allow fish to ingress, and then raised once again to catch another batch of fish on the next ebb. They were highly manoeuvrable and deadly effective in the right situation.

The next obvious development in the history of fishing was the boat. Once man learned to hollow out logs or tie them together to form rafts, he was able to fish into deeper waters. Long-lines were set in deeper water, baited often with shellfish with, as we've seen, cod and haddock being the favoured catch. The same applied to setting creels further out to sea.

But it wasn't until the late eighteenth century, with the Dutch fishing herring within sight of the Scottish coast, added to their development of the drift-net in the eighteenth century, that the Scots sensed a chance to expand her fisheries. Much of this was due to a decline in Dutch fortunes after William III of Orange was placed upon the English throne, thus forming an alliance between the countries which resulted in much of the Dutch trade passing through London. William later passed a decree that the Anglo-Dutch navy should be under English command; that and other matters caused resentment that led eventually to the fourth Anglo-Dutch war (1670–4). The Netherlands' defeat resulted in, among other things, an almost overnight decline in their fishing fortunes.

British government intervention to build a herring fishery soon followed, with Committees being set up in the late 1790s to report on the state of this fishery, and to advise. With the formation of the British Fisheries Society to promote the fishery, it was Wick alone on the east coast that gained, with the building of an enlarged harbour as a model fishing village. Named Pultneytown after the director of the British Fisheries Society, Sir William Pultney, both village and harbour were designed by Thomas Telford and the harbour was completed by 1810. With 200 boats working the herring prior to any work, trade increased to such an extent that a further enlargement of the harbour became necessary in 1824. The Scottish herring fishery expanded quickly to surpass anything seen previously and within a couple of decades Wick was the herring capital of Europe. This continued until the 1970s, when over-fishing led to a collapse of herring stocks and a closure of the fishery for several years. But alongside this seasonal fishery, other fisheries such as that for demersal species, the salmon, lobsters and crabs, and latterly the prawn fishery, flourished as the herring fishery declined.

Line Fishing

Right: Line fishing simply consisted of the setting of baited hooks attached to a line. In its original form, this would have been set on a pole or rod until larger hand-lines were used. With the access into deeper water afforded by boats, lines became longer – hence 'long-lines' – and were set upon the seabed away from the coast. Each consisted of a line with hooks attached at intervals, each of which had to be baited each day. Here fish wives are shelling mussels prior to baiting the lines in the basket, in what is a particularly obvious posed shot.

Below: The *sma* (small) line only differed to the great line in length; both consisted of hooks, each attached to a short length of line about a yard long called a *snood* or *snuid* made of line and horsehair, which in turn were fixed to the main line. Each end of the main line was anchored to the seabed and linked to marker buoys on the surface. A *sma* line consisted of between 600 and 1,000 hooks. Small inshore boats were used to set these close to the shore. The boats in this photograph would all have been used for inshore fishing.

Great lines were much longer and had bigger hooks. The former was used to catch white fish such as haddock, codling and whiting, while the great line caught big cod, ling, halibut, saithe and turbot, among other species. The *haaf* fishery of Shetland was particularly renowned for its hardy types who camped in remote fishing stations over the summer to take their *sixareens*, six-oared open vessels, up to 50 miles offshore to set their great lines. The fish was brought back after a long day at sea and gutted, salted and dried ashore. This photograph shows the fishing station at Fethaland, Shetland, in the 1880s.

The great line boat *Glenstruan*, A200, built in 1958 at Peterhead, was typical of the boats that fished lines off Iceland and Greenland. These boats stayed away from Aberdeen for two weeks, fishing lines up to 10 miles long with hundreds of hooks in depths from 150–350 fathoms on a rocky seabed where trawling was not possible. Halibut was a prized catch. When she retired from fishing, the *Glenstruan* was renamed the *Dawn Gem* and worked in the oil industry, as she appears here.

John Andersen of Stromness, Orkney, baiting the hooks for his long-lines in 1910. Behind and on the ground in front are early twentieth-century lobster and crab creels. (*Photo: George Ellison*)

Above: Small beach-based fishing communities survived into the late twentieth century. Here at Catterline, Kincardineshire, several small boats are seen in September 1962. From right to left the boats are: *Hopeful, Mascot 2, Fear Not,* a pleasure boat, *Linefall* and *Mascot.* The boat drawn up on the hard is an old small fifie called *Barbara* that was retired at that time, replaced by *Hopeful.* The salmon coble ME117 was also working from here at that time, and the poles on the left are for the bag-nets used. The two black and one terracotta coloured sheds are Seagull outboard stores while the orange boxes behind the car are for packing partans (crabs). (*Photo: Edward Valentine*)

Right: Buoys were used to support either end of the line. This photograph shows a buoy made out of the skin of a dog, called a dogbuoy, exhibited in the Unst Boat Haven in Shetland. It was usually stray dogs that were killed for the purpose although other animals such as sheep and goats were also used. Buckie men breed their own dogs for the purpose while Fraserburgh fishers used bullock bladders.

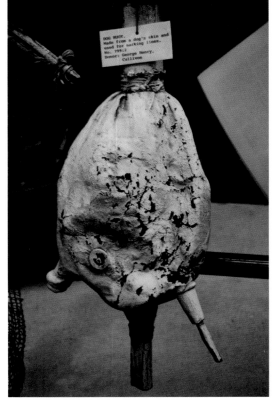

9

Creels

Lobster and crab fishing is an age-old occupation all over the British coast. In its simplest form, individual baited pots – creels – are dropped upon the seabed, attached to a buoy at the surface. Creels were traditionally made from withies and different parts of the country used different shapes of creel. However, by the middle of the twentieth century most were being made with netting around a metal frame or, in some instances, polyethylene pipe. Here, various small Aberdeen-registered inshore boats are pulled up the beach at Catterline in the late 1990s and judging by the number of creels around, many were catching lobsters and crabs.

Small inshore boats have always been the favoured craft of creel fishermen for obvious reasons such as cost, one or two only required and ability to reach areas that large craft could not navigate in. Double-ended creel boats worked under sail before motorisation of the fleets took place in the early twentieth century. However, these small craft adapted well to having engines fitted. Boatbuilders such as Smith & Hutton and Millers of St Monans, both companies in Fife, continued building small creel boats into the second half of that century. This boat is *Quest IV*, built by Millers as *Harriet II* and photographed here in 2001 at Gourdon.

However, today most creel boats are transom sterned, to allow the creels to be shot over the stern. The vessel on the right of this photograph is *Comely III* and belonged to the Reilly brothers, Raymond and Graeme, who fish out of Pittenweem although they live further along the coast at Crail. The boat, like many, is dual purpose so that they can trawl outside of the shellfish season.

Creels being shot over the stern. Instead of laying individual creels out, trains of them are laid along the seabed to save time. The number of creels depends on the size of the boat and how many it can hold at any one time. This boat, *Lena*, belonging to Tom Meldrum of St Andrews, was capable of setting trains of ten creels; the rope leading from the creel already set can be seen as the next creel is about to shoot off the end of the shute.

Tom Meldrum is lifting a creel using his hauler, mounted on the starboard side of the boat. Tom works alone so the controls are at hand – with two crew, the engine and hauler are both operated from the wheelhouse – and each creel in the train is brought aboard, any catch removed and the creel is rebaited and placed on the shute before another creel is hauled. The advantage of these rectangular creels over the round traditional inkwell pots is easy to see as the former stack more neatly and take up less room than the round creels.

11

Salmon

Other than catching salmon in the river *cruives* as described above, there are three other methods: in a drift-net out around the mouth of various rivers to catch them before they enter the river, in stake-nets set along the coast to catch them as they swim parallel to that coast, and in seine-nets within the confines of the river. Here the men of a particular seine of the River Tweed pose along with their gear, which comprises a net, a boat and a capstan, though many seines do not have the luxury of a capstan. Each seine pays an annual licence and operates in the same place on the river as there is a limit to the number of seines on each river. Today most seining is no longer carried on, though there are exceptions.

A good view of the seine net – if not somewhat posed – as it is being hauled in by the men on the shore. However, to set the net, it is rowed out from the riverbank using the boat, sweeping the net around in a circle as the current takes it upstream. We shall look at this fishery in more detail in Book 3 of this series.

The boat is the typical salmon seine-net boat of the east coast. Salmon cobles, as they are called, are the cousins of the Yorkshire and Northumberland larger cobles which we shall see in Book 5. They are rowed by one man while the net sits on the stern platform so that it is shot simply by propelling the boat forward. In the background, the men are hauling in the net.

Another larger salmon coble, seen at Balintore in 1999. Rather than working in the relative calmness of rivers, this one is designed to work offshore. It has an engine rather than sails and a characteristic up-turned prow, copying types seen in Scandinavia. Some say they evolved from the skin boats such as the currach which are known to have worked off parts of the Scottish coast. Working, as the fishermen do, near the coast, the bow copes better with any surf while the crew work.

Stake nets could still be seen until recently. This one was working about ten years ago along the coast north of Montrose. However, these days they are deemed not to be very effective. Although this one is simply a funnel-shaped net, some were extremely intricate and complicated systems of chambers within chambers, designed to confuse the fish and prevent their escape. They would be emptied on the ebb.

Herring

Top: Traditionally, herring were caught in drift-nets set out at night in long trains. These are set so that they float near to the surface of the sea, buoyed at both ends with corks along the top rope and weights along the bottom so that they hang vertically. They stretch down several fathoms and drift with the tide. The sailing boats would sail out in the evening and shoot their nets and then lie to, their nets attached to the bow of the boat. They would be checked regularly by hauling in the first net to see if there were any herring, and then shot again if empty. Once the hauling process did begin, they were hauled in by hand, an operation that might take hours. The boats would race home to unload, the crew catching a few hours of sleep before returning fishing that evening. Here the net is being hauled up over the port side of *Refleurir*, the herring being shaken out once they are aboard.

Middle: Modern pelagic trawlers use methods that entrap huge shoals of fish. Here the net from pair-trawling is being hauled alongside the vessel. These nets are capable of holding a hundred tons of fish and this is then pumped aboard rather than hauling the net in. They are then fed into refrigerated saltwater tanks aboard the boat, with subsequent hauls being added, before the vessel returns to port and the whole catch is then pumped into the processing factory where, in most cases, it is squashed and squelched into fish meal, with the oils being first

removed. Fish meal supports the aquaculture industry in an inefficient way, in that it has been calculated that five kg of fresh fish produce one kg of farmed fish. (*Photo: David Linkie*)

Bottom: Nevertheless, pair trawling using large pelagic trawlers is not without its dangers too. Here the two vessels come close together in calm seas to pass the rope from one end of the net across. With one net between two boats, it is necessary to pass this rope across twice, once before the net is shot and again once trawling has been completed and the net is about to be hauled alongside by one boat. In rough seas, getting close enough to physically throw a small line across can be perilous without extreme care as the waves are capable of throwing the stern of one of the vessels into the other to cause a collision.

Prawns

In 2001, the author joined the crew of the *St Adrian* out of Pittenweem on a typical day's voyage of prawn fishing. The 12.94 metre *St Adrian*, KY360, was built in Gravesend in 1970 and was, when new, highly modern, boasting a low wheelhouse with a sheltered working deck and a large drum winch situated just behind the wheelhouse. A sequence of photos follows.

This boat has two crew and the winch is controlled by the helmsman. Here the net is being prepared for another shot.

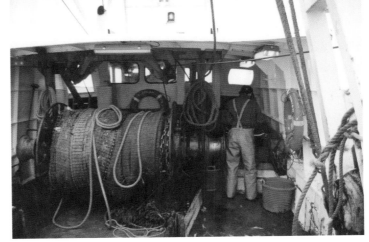

The net is starting to be hauled in.

The cod end is lifted high, prior to being opened.

The catch is lying on the deck and consists of all manner of small fish, crustaceans and sea life, much of which is discarded over the side as either useless or unquotaed.

The prawns are sorted and put into the various boxes, depending on their size and quality. Small prawns have their tails removed at sea and will go as scampi, while the larger prawns will be sold as langoustines.

Seine-Net

At the onset of war in Europe in 1914, the huge markets Britain had forged in Eastern Europe – in Prussia and Russia for example – disappeared overnight and never really survived past 1920. Fortunately, in 1920, the seine-net method of fishing for white fish at sea arrived in Scotland after Scottish fishers had seen Danish 'snibbies' from Esbjerg landing into Grimsby. By the autumn of that year Aberdeen fishermen were using seine nets, which caused outcries in some fishing communities. Trawling legislation banned fishing with a trawl in areas near the coast, thus protecting the line and creel fishermen, but there was no control for the seine net. The boat *Marigold*, seen here, owned by John Campbell of Lossiemouth and built by William Wood & Sons in 1927, was the first purpose-built seiner in Scotland.

The seine net evolved in the waters of Denmark in the middle of the nineteenth century. The Scots developed their own version – fly-dragging – which allows faster swimming fish such as cod, haddock and whiting to be caught. Whereas the Danes had anchored the net, the Scots didn't and the net is attached to two very long ropes around two miles long. The net is deployed in a triangular fashion using a buoy to hold one end while the boat sweeps around and steams back to the buoy and complete the set. Both ropes are then winched in, with the boat moving forward slowly. Speed increases as the net gets close to herd the fish into it before it is closed and they are completely encircled. Here the seiner *Verbena*, INS90, is coming into Lossiemouth.

Fishing Boats

Early fishing boats were simple open boats, often constructed by the fishermen themselves using flimsy wood and poor nails. Up to the 1880s, there were two distinct types in use on the mainland east coast – the scaffie to the north and fifie to the south. Shetland and Orkney types were slightly different and open boats remained in use longer among the islands, though some fishermen later adopted the smack-rigged fifie. By the 1860s many larger Scottish boats were being decked over, although this initially attracted the view that this was at the cost of valuable cargo space for fish. By the 1880s the Zulu was introduced in the Moray Firth, a large and powerful sailing fishing boat which was considered a hybrid of both scaffie and fifie. Small inshore boats were, as often as not, simply smaller versions of the larger boats (hence fifie yawl, Zulu skiff etc). However, steam was beginning to affect the fleets and within a decade of the new century motorisation of the fleets had begun. Fishing boat design dramatically altered to suit these engine units and gave birth to the Motorised Fishing Vessel (MFV) that Scotland became renowned for. Further improvements in the late twentieth century came about as safety and comfort demands were taken into account, with such innovations as covered working decks, cabins for crew and wheelhouses bristling in electronic gear. Fish-finding became more computer-controlled, as did the manipulation of the fishing gear.

Sailing Fishing Boats

The profile of the scaffie demonstrated a pronounced curved but deep forefoot and a sloping sternpost. It was generally rigged with one dipping lug sail although a few of the larger versions did have two masts. They were designed specifically for the herring fishing and were an improvement on the very basic open fishing boats in use prior to about 1820. They tended to be favoured by the fishermen from the north, between Duncansby Head and the Moray Firth. Here, several sit dried out in the harbour of Portknockie, the outer one sitting next to a small fifie. Note the bows of larger boats, probably fifies, lying off to the right in deeper water.

Right: The small skiff *Gratitude* was built at Portknockie in 1896. Here is a replica of that vessel, the *Obair Na Ghaol*, built in the traditional way over 16 months by Alex Slater and Sinclair Young for Portsoy Maritime Heritage. She was then launched at the Scottish Traditional Boat Festival in 1996 and has been a constant visitor to that festival.

Below: Fifies were recognisable by their upright stems and only very slightly raked sterns and were said to have originated on the Fife coast, where they were influenced by Dutch boats. Their use spread northwards towards the Moray Firth. By the 1860s most fifies were decked over, although some older boats remained either open or half decked. Here a crew can be seen working on the deck of a fifie with its two dipping lugsails rigged to dry. This is a small fifie at about 40 feet in length. The men appear to be repairing the nets and ropes.

Above: The *Morning Star*, KY190, here is a prime example of a large fifie, in excess of 70 feet in length. The mainmast would be about 60 feet long, with a huge dipping lugsail that would take the seven crew to lower on its yard, and dip around the mast, when tacking. The letters KY are the registration letters for the port of Kirkcaldy, each port around Britain having its own letters which allow that port to be identified. The numbers are unique to that vessel from this port. All fishing boats have been registered in this way since the Sea Fisheries Act of 1843. The same letters and numbers should be clearly displayed upon the mainsail as well.

Left: The fifie *Isabella Fortuna* is shown here in the harbour at Wick in 2011. She was built as the *Isabella* by James Weir of Arbroath in 1890 for the line and drift-net fishery. At 45 feet, she is a small fifie. She fished under the ownership of the Smith family of Arbroath for 86 years. In 1919 a Kelvin 15hp engine was fitted but by 1928 a more powerful Kelvin 44hp K2 had to be fitted for the seine-net fishery. Four years later, a larger K3 66hp was fitted and her name was changed to *Fortuna*. When the Smiths retired in 1976, she was bought by Hobson Rankin, who restored her. In 1980 she was renamed *Isabella Fortuna* and in 1997 she was handed over to the Wick Society, whose volunteers operate her during the summer months to various sea-based events.

Here four boats are lined up in Whitehills harbour on the Moray Firth. Three at least are registered at Banff and have small pill-box wheelhouses, which suggests that the photo is from the early twentieth century. Steam capstans are visible and judging from the lack of sails, it is possible they have been motorised. The nets lie drying over the main yard. BF176 is the *Abstainer*, a Zulu built by Stevenson & Asher of Banff in 1899.

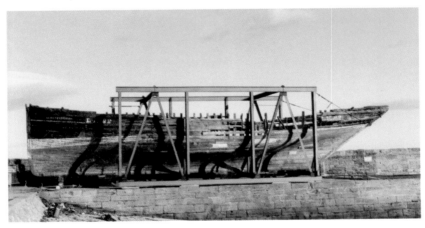

The *Zulu Research*, LK62, was built as the *Heather Bell*, BF1206, by W. & G. Stephens of Banff in 1903 and skippered under Alexander Paterson, and is typical of a large first-class, two-masted Zulu. She fished out of Macduff. She was sold in 1912 to Rosehearty, where she was registered as FR498 and was motorised in 1917. After another two owners she went to Shetland in 1935, where she was renamed *Research* and registered as LK62. She fished herring until being requisitioned for war duties, after which she resumed fishing in 1946. In 1979 the Scottish Fisheries Museum salvaged her and returned her to St Monans, where some major repair work was carried out. In 2000 she became a static exhibition inside the museum. In this photo the *Research* is shown in her cradle, into which she was placed in 1996, prior to being sited in the new boat gallery.

Above left: The 82-foot *Muirneag*, SY486, was one of the last Zulus built by McIntosh of Portessie and Ianstown, who had been building fishing boats since around 1830. Messrs McIntosh had already launched the first Buckie-built steam drifter, *Frigate Bird*, in 1900 but in 1903 were still building mainly Zulus, six being launched in 1903, each taking about eight weeks to build, and costing £500 including spars. A further four were launched in 1904 before they presumably adapted their yards to the production of steam drifters. The *Muirneag* – 'Darling Girl' in Gaelic and also the name of a prominent hill near Stornoway – was ordered by Alexander 'Sandy' MacLeod in 1903; it was his second McIntosh-built Zulu and he had a reputation as a fine fisherman and mariner, his first Zulu, *Caberfeidh*, being the first Stornoway Zulu to go the English herring fishing. This reputation was enhanced as he fished *Muirneag* continuously up to the outbreak of the Second World War, refusing to convert her to engine power and thus being the last British herring drifter to fish under sail power alone. However, he retired in 1945 and the boat was sold in 1947 by public auction in Stornoway for £50. She was dismantled and cut up for fencing posts, although local dental mechanic George MacLeod took her measurements, from which drawings were made by Harold Underhill.

Above right: Here, a smaller Zulu yawl, *Ivy*, KY18, is photographed with the crew posed aboard. These smaller boats were used principally for inshore fishing of lines. Rigged with one dipping lug, they did not need a crew of six, whereas the larger boats, with their huge powerful dipping lugsails, did need the full crew complement.

Small Inshore Craft

A small fifie yawl, *Alexander*, 600A, owned by George Woods, poses with three men and three children aboard, thought to be possibly at Portlethen, outside Aberdeen. This size of boat, with its single dipping lug and oars, would work lines close to the shore. Several more similar vessels lie out of the water behind.

The small fifie *Robina Inglis*, LH179, is seen lying in Newhaven harbour, Edinburgh, in about 2002. She was built at Newhaven in 1923 by Brown & Allan for fisherman Tam Wilson. She worked from Eyemouth for some time and was later owned by the last working fisherman in Newhaven, David Brand, when this photograph was taken. Newhaven is said to be one of the oldest fishing harbours on the Firth of Forth. In 1875 there were more than 400 fishermen working from the harbour in at least 100 boats. Today, this boat is owned by the Berwickshire Maritime Trust and is based at Eyemouth; she has been renamed *Good Hope*, her name during her original time at that port.

Top: This unnamed boat was photographed in about 1997 and is an example of a Fraserburgh *yole*, a type of small inshore boat built specifically in Fraserburgh for the inshore fishing of creels and sma' lines. Several still survive in varying states of repair while at least one is back in sailing condition, this being the *Robin*, owned by Bertie Gillespie of Orkney. The boat was bought in 1982 and Bertie fished for creels out of Orkney until he converted the boat back to sail in about 2003.

Middle: The Fraserburgh yole *Progress*, PD212, leaving Peterhead harbour. At first these boats were open, though most were decked over by the late eighteenth century. This boat survives in Cornwall. It would appear she is working the seine net here, judging by the coils of rope on the deck and the Beccles line coiler.

Bottom: Restless Wave, FR2, another yole, steaming out of Fraserbugh, heading out to the sma' line fishing. Many of these small boats were built by Tommy Summers at his Fraserburgh yard.

Top: This is a typical small fifie which has been motorised by chopping away part of the sternpost and the addition of a steel plate to build out the keel. Here *Primrose*, PD372, was photographed at Boddam in about 1997. She is a typical creel boat of unknown vintage, though her hauler is still visible. A photograph of 2010 shows her rotting against a bank, her days being over, a situation typical for many of these old fishing boats. Thankfully, enthusiasts have saved some by restoring them to their former state.

Middle: Another example of a traditional type being motorised. This boat, photographed in 1997 at Lybster, is thought to be the Stroma yole *Pandora*, built prior to 1889 for George Robertson. The rudder has been swapped for a steel version and a small cuddy built over the bow. Stroma yoles were similar in design to Orkney yoles, as we shall see, though they were also related to the generalised 'Wick' boat.

Bottom: A Dysart yawl, *Girl Linda*, sitting atop the quay at Dysart in Fife, on the north coast of the Firth of Forth. Dysart, an ancient port associated with the export of salt and coal, also had a vibrant fishing industry. It is said that salt, coal, beer and cured fish were exported to Holland in return for 'cart wheels and delft-ware, kegs of Hollands and pipes of Rhenish' and because of the

amount imported, the town, now a suburb of Kirkcaldy, became known as Little Holland. Dysart yawls were small open boats with little sheer, used as a model for the St Ayles skiffs being built.

By 2010, it would appear that *Girl Linda* has been restored and converted, although it is by no means certain that this is the same boat. They do appear very similar. However, it does illustrate how life can be drawn back into some of these old boats with the right determination and finance.

Two small creel boats at Anstruther in 1997. These were typical of the creel boats built by Millers of St Monans and Smith & Hutton of Anstruther.

Two typical small motorised Wick boats moored up at the small harbour of Keiss in 1998. These double-enders, like the Stroma yole in 2/15, display similar characteristics to the yoles of Orkney.

Top: Sea Ranger, WK495, photographed in Wick in 1997, shows how flat the floors of these boats are in section. In the Washington Report of 1849, he described the 'Buckie or Moray Firth boat' as being one of the most seaworthy of all Scottish fishing boats and was sometimes known as a 'scaith'. The Wick boat, much more upright in profile and fuller in section, was deemed fairly unseaworthy! However, today's Wick boat has a stronger semblance to the Moray Firth boat than the Wick boat, which suggests the fishermen looked to the south to develop their boat on similar lines to the way the scaffie evolved.

Middle and bottom: Two views of a typical salmon coble. The *Sandra* of Montrose was photographed at the Lunan Bay salmon fishing station of Fishtown of Usan in 2000. These flat-bottomed boats work well when launching into waves.

27

Motorised Traditional Sailing Types

Top: Sovereign, BF1289, was one of the first motor powered boats on the Moray Firth. She had a motor installed in 1909 and is seen here leaving Macduff harbour soon after, judging from the lack of a wheelhouse or capstan and the fact that a mizzen sail is set. The amount of nets on deck shows just how many nets these herring boats shot each night during the herring season. Boats such as this would fish throughout much of the year, following the herring down from Shetland to the East Anglian autumnal fishery. Some boats travelled over to the west coast to fish during their season.

Middle: The unnamed motorised fifie ML237 from Methil powering out to sea from Pittenween. This photograph shows the sheer power of these motorised vessels. She is equipped with a wheelhouse, engine room housing, small punt and a steam capstan but still retains the mizzen sail, which suggests that this photo was taken in the 1930s. Seven crew are visible and presumably the skipper is in the wheelhouse.

Bottom: The motorised fifie *Brighter Dawn*, KY656, at St Monans in around 1934. Again she is fitted with a wheelhouse and a steam capstan. The steam capstan enabled boats to set more nets than was previously possible when they were hauled by hand, a tedious and arduous task in calm water, never mind rough seas. Sometimes the hauling process would take hours. These big fifies were deemed suitable to receive engines and so many were motorised. The Zulus, with their sloping sternposts, were not nearly as suitable, though some indeed were.

Three views of the
45-foot motor fifie
Betty Yorke, BF434,
built by the Macduff
Engineering Co. in
1938 for fishing with
the seine-net. The first
photograph shows
her as she was after
launching. Note the
outboard rudder, a
characteristic of the
sailing boats that
didn't change until the
1940s and 1950s. The
second photo shows
her in Arbroath in
about 1997, when she
was being used as a
liveaboard. The port
rail has obviously been
damaged. I believe
she then sank and the
third photo then shows
her while some more
conversion work was
underway.

The Zulu *Violet*, FR451, was built by James Noble of Fraserburgh in 1911 for Alexander Grieve Stephen at a cost of £90. The boat fished for herring during the season, and with sma' lines in winter. Today, after a lengthy restoration, she survives as a private boat based in Martha's Vineyard in the US.

This photograph shows the sister boat of the *Violet*, the Zulu *Vesper*, FR453, also built by James Noble in 1911. She was owned then by George Noble and John Buchan. Both were motorised soon after being launched.

This page and next page above: These photographs show *Vesper* in various states during her lifetime. She ended her days out of the water in Buckie while the owner attempted to obtain financial help to restore her. Sadly, he failed and she was later dismantled.

The Zulu *Evangeline* ashore at Anstruther in 1998. It has been suggested that this boat had been built to replace an earlier 1896-built Zulu *Evangeline*, BF1952, from Portknockie, that was lost with all the crew during a ferocious gale in 1905. This *Evangeline* was destroyed by fire soon after the photo was taken.

 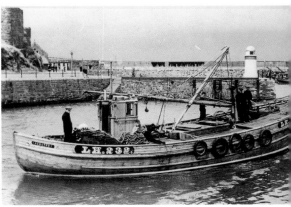

Motorised Fishing Vessels (MFVs)

Above left: The *Efficient*, FR242, was built by J. & G. Forbes of Sandhaven, near Fraserburgh, in 1931 as a herring drifter for the Ritchie family. She was engined with a Petter Atomic diesel of 160hp and her owner was proudly photographed standing in the engineroom in Petter's engine brochure. At 82 feet in length, *Efficient* was a large boat thought able to compete with the steam drifters but, even though she line-fished in winter, six years after her launch she was sold for £393 15s to William Stevenson & Sons of Newlyn and was converted for trawling. She was requisitioned in 1941, and used initially to help ferry POW escapees from Norway to Scotland. She later ferried King George VI among the Scottish isles during a royal visit. She returned to Newlyn after the war, was re-engined with a Lister Blackstone and, at the same time, renamed *Excellent* after the original lugger owned by William Stevenson until 1888. She was registered as PZ513 and she survives in Newlyn today. (*Photo: Billy Stevenson*)

Above right: *Achates*, LH232, was built by Walter Reekie of St Monans in 1949 as a ring-net boat. She fished out of Fisherrow and was a typical varnished ringer, built without bulwarks to ease the hauling of the ring-net. This method of catching herring was developed from early uses of a seine-net. Basically, two boats are employed to shoot the ring-net around a shoal of herring in relatively sheltered waters. The method was pioneered on Loch Fyne on the west coast and we shall discover more about the ring-net in Book 2. However, many ring-net boats were built by boatbuilders on the east coast and the boats sailed to the west coast to fish, with some ring-netting taking place in sheltered parts of the Moray Firth. Here, she is seen leaving Peel in the Isle of Man, where ring-nets were also worked.

Top: The *Rachel Douglas*, BK231, was built by William Weatherhead & Co. in their Eyemouth yard in 1947. Although registered in Scotland (BK is Berwick-upon-Tweed), she worked out of Seahouses, fishing with a ring-net and seine-net at different times of the year and also working lobster pots at times. She was fitted with a Kelvin 44hp engine at launch though this was replaced with an 88hp version soon after. When she sold in 1963, the boat moved to nearby St Abbs, where she fished until 2002. She was then acquired by the North East Maritime Trust and restored by South Shields boatbuilder Fred Cowell and today the boat is in their collection and is kept in Newcastle and sailed to various maritime festivals along that coast. She is shown in about 1950 in this photograph, coming into Seahouses with a full load of herring.

Middle: The *Rachel Douglas*, now painted blue, is seen on the right of this photograph. On the left is the former seine-net boat *Favourite*, BK11, also owned by the North East Maritime Trust. She was built by Walter Reekie of St Monans in 1947 and initially worked out of Castletown, Isle of Man, as *Margaret Anna*, CT101. Within a few years she was at Amble and, in the late 1960s, she was renamed *Favourite* and, like the *Rachel Douglas*, she then worked out of Seahouses for twenty years, seining and potting before also moving to St Abbs, registered as LH149, where she trawled. When she left the fishing she was restored by the NEMT and is also kept in Newcastle's St. Peter's Basin. (*Photo: Mike Craine*)

Bottom: The launch of a new boat was always a celebratory affair and a time when all the family of the new owner would dress up and, after the boat was launched, join the builder for a slap-up meal somewhere. Here, the new owner of *Floriage*, A56, John Murray and his wife are about to officially launch the boat, the wife holding the champagne. (*Photo: Ian Murray*)

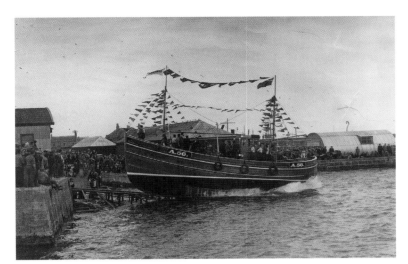

The launch of the *Floriage* at Gerrard's Yard, Arbroath, in 1955. (*Photo: Ian Murray*)

Floriage arriving in Aberdeen for the first time after her launch. The Torry Fishery Research Station is in the background. (*Photo: Ian Murray*)

Floriage working out of Lowestoft after she had been sold to David Wells of Great Yarmouth in 1976, after John Murray came ashore. (*Photo: Ian Murray*)

Floriage working out of Balbriggan (SO is Sligo). (*Photo: Ian Murray*)

Green Pastures, LH424, was built by Millers of St Monans, the premises of which can be seen in the background in this photograph. Millers as a boatbuilding firm began life in 1779 in St Monans and continued until its closure in the 1990s. During that period they built all manner of fishing boats and gained a reputation for producing some of the finest boats ever seen. They also built several pioneering boats such as the two ring-net boats *Falcon* and *Frigate Bird* for the Campbeltown fleet in 1922. These motorised ringers were of a new, totally different design to the earlier skiffs, and had been influenced by Scandinavian boats. Millers also had a yard in Anstruther where, in 1905, they built and launched the *Pioneer*, the first motorised 72-foot fifie. Though deemed unsuccessful, within a few years most boats were being fitted with engines.

Another lovely varnished herring boat, the *Jacquamar*, FR95. She was built in Fraserburgh in 1971 and neighboured the *Dewy Rose* at the ring-net. However, by the 1990s she had had a new wheelhouse and the hull had been painted red. It seems that she ended her days as a bare hull in Ireland. Note the transom stern.

Two very different boats alongside. The canoe-sterned *Zephyr*, INS6, was built by Weatherhead & Blackie of Dunbar in 1969 as a ring-net boat and was originally varnished. She was Avoch-owned. The other vessel, with the cruiser stern, *Incentive*, INS155, was Danish-built in 1966. *Zephyr* had the distinction of being the last boat to land herring into Inverness in 1984. By the 1990s she was trawling for prawns.

The 1965 Banff-built trawler *Alex Watt*, INS163, fished out of Lossiemouth for a number of years. Here she is seen with the side shelter deck built in steel and fitted against the original wheelhouse. She is a stern trawler, as seen by the two otter boards fitted either side. These are lowered into the water with the trawl net behind and act to hold the mouth of the net open.

Good Fellowship, BK172, was built by Gerrard Brothers of Arbroath in 1960 as the dual purpose trawler/seine-netter *Sunset*, A430. She has had a new wheelhouse complete with side shelter deck in this photograph. Today she still fishes and has even more modernisation, including a high structure on top of her wheelhouse holding an array of aerials and radar gear. There's also a new power block on its own gantry, mounted behind the wheelhouse. Power blocks were first introduced onto fishing boats to haul in purse-seine nets. It was invented by a Croatian fisherman working in California and it soon became popular on all sorts of vessels as the technology improved in the latter part of the twentieth century.

The Eyemouth-based trawler *Rebecca*, LH11 (was *Heatherbelle VI*, LH272), built by Herd & Mackenzie in 1987 and photographed in Eyemouth in 2008. Although her hull is wood, she now has a shelter deck extending forward of the wheelhouse right to the bow. Again, there is a power block fitted.

Realm, BCK 183, is seen at Eyemouth in 2008. *Realm* was built in Denmark in 1968 as a seine-net boat and is similar in hull shape to the hundreds of these built on Denmark's west coast, especially working out of Esbjerg. Many were introduced into the Grimsby fleet and were called 'snibbies' while a few made it to Scotland. They are renowned for the pretty sheer line and cruiser stern and originally would have single-cylinder diesel engines such as Wickmann. However, this boat has had a Gardner engine installed and a full shelter deck, power block and otter boards fitted for trawling.

The 1983 Macduff-built *Watchful*, BF107, is seen trawling in 2008. The winch drum can be seen at the stern, with the power block just forward. She has a Kelvin 495hp engine which provides plenty of power to trawl a large net. Again, her foredeck is enclosed for the safety and comfort of her crew.

Left: The large drum winch with net and gear aboard an Eyemouth stern trawler. The large rubber wheels called bobbins allow the net to run along the seabed while the buoys hold the upper mouth up. The otter boards hold the mouth of the net apart. Obstacles on the seabed can be very dangerous, snagging the net and pulling the stern down so that the sea floods in. Many boats have capsized and sunk in this way.

Below: Wheelhouses are an integral part of a fishing boat and contribute immensely to the beauty and variation in Scottish vessels. Here is a 1950s wood and steel wheelhouse aboard a gill-netter, chosen because of its anonymity although someone will surely recognise the boat. However, wheelhouses come in all sorts of sizes, shape and materials and many are recognisable as being the product of a specific boatbuilder.

Steel Vessels

Above: The stern trawler *Radiant*, PD298, in Peterhead in early January 2002. She had been built the previous year in Asturias, Spain, and was a twin trawl boat, meaning that two nets side by side are shot during the fishing process. At the time it was thought that the boat was both ugly and somewhat top heavy. The boat has a huge bulbous bow, though she does have a deep draught. However, while fishing 45 miles north-west of the Isle of Lewis only a few months later (April), the port trawl snagged a seabed obstruction and became fast, eventually causing the boat to capsize and founder. One crew member was lost.

Right: The net of the 1980 Norwegian-built pelagic purser-trawler *Challenge*, FR77 (ex *Zephyr*), is seen through the gantry. The massive nets are wound onto the drum of the winch, powered by its 2,200 hp Blackstone engine. At 56.70 metres in length, these boats are formidable tools, using the latest electronic gear to locate fish, and powerful systems to entrap huge shoals.

The end of the trawl net, where the fish are finally forced into, is called the cod end. It is the last bit to be winched aboard. Once it is, and it is suspended over the deck, the bottom of it is untied to allow the fish to fall out. Only then can it be sorted into the various species, with what is either not wanted, or for which the boat has no fishing quota, is thrown over the side. This is referred to as the discards.

The net being hauled aboard. This photograph simply shows the nature of the net as it is being winched aboard.

The cod end being swung over the side of a boat using the power block. The crew look on with anticipation, wondering how much good fish they have landed.

Shetland and Orkney Boat Types

A Shetland fishing sixareen (or six-oared boat) being used to ferry working ponies in Uyeasound in about 1905. These boats, descended from the Viking boats of Scandinavia and often brought over from Norway in kit form to be built on Shetland, were substantial carriers. Often referred to as the *haaf* boats (deep sea boats), they were long-line boats that went out to fish for cod, haddock and ling up to 50 miles offshore. They were traditionally rigged with a squaresail, though they evolved into a dipping lugsail in time. (*Photo: James M. Smith*)

A smaller fourereen (or four-oared boat), the *Maid of the Mist*, on the beach at North Haa on Melby. These boats were used for inshore fishing the winter haddock and seldom ventured more than 20 miles away from the islands. Like the bigger sixareens, they set a squaresail which later became a dipping lug. (*Photo: J. D. Rattar*)

Ness yoals in their nousts at Spiggle, Dunrossness, in 1905. These boats generally worked the waters around the southern part of Shetland with hand lines and were regarded as the most elegant of the Shetland types. They often worked the fast moving tidal waters between Sumburgh Head and Fitful Head, where a swift, reliable vessel was vital to ensure that the boat was not thrown onto one of the many hazards. Again, these were squaresail rigged until lug became favoured.

Shetland boatbuilder Tommy Isbister rows one of his Ness yoals. Tommy started building boats in the early 1950s and also spent many years away at sea, fishing. However, when yoals were taken up to be used as racing boats, many new yoals were built. Between 1994 and 2002 he in fact built eleven yoals. Today, Shetland has any number of yoals which race in the same way as the Cornish (and others now) race their gigs. However, traditionalists like Tommy keep to the same design while others have tweaked to get more efficiency when racing. But to Tommy, as it is to the hundreds of fishermen lying in their graves, 'a yoal is a yoal, and nothing else is'! They regarded boats of differing dimensions not to be yoals.

The Unst Boat Haven on the most northerly island of Unst is, as it says, a haven for these old boats. Housed in a purpose-made building, there are some dozen or more traditional Shetland boat types, as well as artefacts from the fishing. The replica sixareen *Far Haaf*, built by the museum's founder Duncan Sandison, also sits outside.

Unst boatbuilder Willie Mouat in his shed discussing his new Shetland boat. Willie worked on many of the boats in the museum and also helped build the *Far Haaf*. He continues building today and has recently (2012) launched his latest racing yoal, *Siri*, which was gifted to the Lerwick Boating Club by Arch Henderson LLP, a Scottish-based civil and structural consulting engineer.

The Shetland fifie *Swan*, LK243, sailing into Stromness, Orkney, in 2002. Both the Shetland and Orkney fishermen built fifies to fish the herring. However, instead of dipping lugsails, they favoured the gaff rig. The *Swan* was built by Hay & Co. of Lerwick in 1900 and was originally lug rigged, this being re-rigged in 1908 as gaff. She would drift for herring in the season and long-line for white fish in winter. In the late 1940s, she worked the seine net but within a decade she had been retired and eventually ended up in Hartlepool as a derelict and sunken houseboat. She was bought and returned to Shetland and the Swan Trust was formed. In the 1990s she underwent a six-year restoration project and now takes parties sailing each summer as part of her sail training duties.

The seine-net boat *Milky Way*, LK106, on the slip at Scalloway. She was built in 1934 in Fraserburgh and represents one of the original seine-net boats. She fished up to the end of the twentieth century, after which she was sold and was last heard of in the Canary Islands.

Another Shetland seiner is the *Pilot Us*, LK271. She was built by Forbes of Sandhaven in 1931 and worked from Shetland after 1946. Based at Scalloway, she fished lines and the seine-net throughout her career, which ended in 2000. She was bought by the Shetland Islands Council and is kept at Scalloway. (*Photo: Mike Craine*)

The seiner *Athena*, LK237, was built by Jones Buckie in 1972 and named after the Greek goddess, the beautiful one. She fished out of Lerwick for a number of years. (*Photo: Mike Craine*)

However, like hundreds of other boats in the British fleet, *Athena* was deemed surplus to requirements in 2010 and decommissioned. She was scrapped and broken up in Macduff in 2011 as part of the 'legalised vandalism' exercised by the British government as part of the policy to reduce over-fishing. Hundreds of perfectly sound boats, built by skilled boatwrights to last generations, have been destroyed in this way, destroying at the same time much of the country's fishing heritage. No one has ever able to explain why these boats could not be put to further use outside of fishing. The bureaucrats found it easier just to pay to have them reduced to rubble.

The 64.4 metre pelagic trawler *Antares*, LK419, seen here at Lerwick, was built in 1996 in Norway. Many of these trawlers, which are generally based in Lerwick or Fraserburgh, were built in Norway.

Above: One of the largest to date is the *Research*, LK62, which at 70.7 metres and having a 7680hp engine, was built at Flekkefjord, Norway, in 2003. Note the huge otter boards at the stern. (*Photo: Mike Craine*)

Right: Len Wilson standing in the replica Orkney yole *Gremista* that he built in his garage. The yole is rigged with two spritsails in the south isles tradition, while in the north isles they prefer the lug rig. Unsurprisingly, these boats are not dissimilar to the Stroma yoles and share characteristics with the Wick boats.

49

An Orkney yole in its noust at Rackwick, Hoy, in the 1890s. (*Photo: Orkney Library Archive*)

The *Island Queen*, K420, still retains some Orkney characteristics in that the bow is slightly similar to that of a yole, and keeps the clinker building thread. The transom stern is vital for increased engine power and working deck space. However, unlike many of today's modern boats, this one has managed to keep a pretty shape. (*Photo: Mike Craine*)

Steam Vessels

Above left: Steam impacted on the fishing industry in a big way. The first experiments with steam began in the 1850s, when smacks and luggers were towed to the fishing grounds by steam-powered vessels. The next step was to tow the sailing vessels with their trawl gear down until William Purdy of South Shields used his tug as a trawler. By the early 1880s steam-powered trawlers had appeared, with staggering fishing results. For four tons of coal, the boat could steam at nine knots, fish and get home almost before the sailing boats had shot their nets. However, the investment was massive compared to a sailing boat. In Scotland, the first steam trawler was the *Rob Roy*, LH92, launched at Leith in 1882. These early steam trawlers retained the rig, such was the likelihood of breakdown. However, steam in the herring fishery had to wait another 15 years or so as these steam boats were not deemed good for shooting and hauling drift nets. That soon changed and by 1913 there were over 1,800 steam drifters. The first steam drifter in Wick, *Content*, WK54, is shown here.

Above right: A Buckie-registered steam drifter at the herring fishing station at Burravoe, Shetland. Boats assembled in Shetland for the spring herring which sounded the beginning of the North Sea fishery that ended up at East Anglia in the autumn. These drifters were over 80 feet in length, some even over 90. They reflected a huge investment as in the early twentieth century a new boat cost in the region of £2,000 for the boat alone, without the fishing gear.

Top: The *Lyre Bird*, BCK174, represents the typical steam drifter. The nets are shot from the foredeck, which has plenty of space, the wheelhouse being aft of amidships. The engine casing is behind that, with the characteristic high funnel with the registration letters always painted on. This funnel led to their nickname of 'pipe stalkies'. A small punt was always carried aft. Accommodation was aft, with fish holds and net rooms forward. Wooden drifters were common until riveted iron boats gained favour in the early twentieth century, although iron steam trawlers had been around for over a decade.

Middle: The steam trawler *Ben Screel* was wrecked at Girdleness in January 1933, after she missed the entrance to Aberdeen harbour in thick fog. Aberdeen was the main trawling harbour on the east coast and had a fleet of these trawlers, though trawled white fish was landed into many other ports. The first to land in Aberdeen, in 1882, was the converted tug *Toiler*. The following year the first screw-propelled steam trawler was launched from Duthie's Yard in Aberdeen, and by 1914 it was the largest fishing port in Britain. By 1930 there were 270 steam trawlers registered there. However, with the decline in the herring fishery in the middle of the twentieth century, other ports such as Peterhead thrived at Aberdeen's expense. By 1990, Peterhead was Europe's leading white fish port. But by then the design of vessels had changed dramatically.

Bottom: Steam drifters at Buckie. Literally hundreds of these boats were built over a thirty-year period. The amazing thing is that, unlike most of the sailing fishing boats, which have several examples still sailing (except the first class Zulus), and more so in the case of the later motorised fishing boats, only two or three drifters have survived to sail today. (*Photo: Campbell McCutcheon*)

Diesel Trawlers

Above left: Aberdeen received its first diesel trawler, *Star of Scotland*, A425, in 1947, just after the war. The *Craigmillar*, BA303, shown here, was built by Fairmile Construction Co. Ltd of Berwick-upon-Tweed in 1959 and fished out of Aberdeen as A303. She was a typical Fairmile 'pocket trawler', also nicknamed 'Sputniks' after the Russian satellite. She worked out of Aberdeen until 1970, after which she went to Fleetwood and then, in 1979, she was registered in Ballantrae. She was decommissioned in 1998. (*Photo: Mike Craine*)

Above right: The Surrey-based Fairmile Construction Co. Ltd took over the Berwick-upon-Tweed yard in 1953, a few years after Weatherheads left. Although the name Fairmile was well known in naval circles, fishing boat design was new to them. However, the Fair Isle class of trawler was launched in 1956 with the building of their first trawler, *Coral Isle*, SN22. Designed by French naval architect M. Guerault and adapted by Fairmile, these boats replaced older, bigger vessels largely because they had lower running costs, both in terms of fuel and a smaller crew, therefore making them popular with owners. Over the next few years twenty-two of these vessels were built, with five being launched on average in a year. Most went to Aberdeen and Granton, though two did go to England. However, with the cessation of Whitefish Authority grants and loans in 1961, orders disappeared. Fishing boats were not built again at the yard until 1969 with the arrival of the Croan class of seiner-trawlers, followed by the Tynecraft class between 1973 and 1977, though neither were built in very large numbers. Fairmile, however, had already sold the yard in 1972 to Intrepid Marine and in 1979 it closed for the very last time. However, with the loss of Icelandic fishing grounds after the Cod Wars of the 1970s, deep sea trawlers declined. The modern trawler had arrived (see top photo, p.41).

53

Boatbuilding

Two wooden steam drifters on the stocks side by side at Herd & Mackenzie's yard at Findochty, c. 1908. The majority of vessels were built out in the open, as seen here. Two boats could be built quicker. The boat farthest from the camera is well ahead of the nearer, which is only in frame as the far one has been planked up. The sawn timber lengths lie on the far side of the boats. The scaffolding around the boat wouldn't perhaps please any Health and Safety Inspector today!

Unnamed boat being launched at James Miller's yard. Work on the deck and superstructure can be completed while the boat is afloat, allowing another hull to be started on the slipway. Note the crowd which is always present for a new boat launch.

This photograph shows the *Olive Branch*, LH366, going down the ways in 1960. She, like many of these dual purpose ring-net and seine-net boats, was the product of the Weatherhead & Blackie yard. Weatherheads started building in Port Seton in 1880 when William Weatherhead, who had come from his father James's yard in Eyemouth, commenced work. He built mainly fishing boats and by 1885 he had moved to larger premises at nearby Cockenzie. The company survived throughout the first half of the twentieth century but, in 1954, was then bought out by Samuel White of Cowes, Isle of Wight, who continued building under the name of William Weatherhead (1954) Ltd and who were the actual builders of *Olive Branch*.

Boatbuilding at the Stromness yard of James Anderson, a photograph taken from a booklet issued by the Highlands & Islands Development Board describing boatbuilding and boatbuilders in northern Scotland in the 1970s.

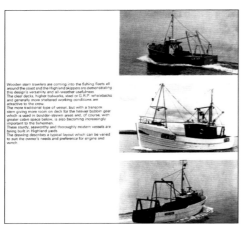

Wooden stern trawlers are coming into the fishing fleets all around the coast and the Highland skippers are demonstrating this design's versatility and all-weather usefulness. The clear decks, higher bulwarks, steel or G.R.P. whalebacks and generally more sheltered working conditions are attractive to the crew. The more traditional type of vessel, but with a transom stern giving more room on deck for the heavier bobbin gear which is used in boulder-strewn areas and, of course, with greater cabin space below, is also becoming increasingly important to the fishermen. These sturdy, seaworthy and thoroughly modern vessels are being built in Highland yards. The drawing describes a typical layout which can be varied to suit the owner's needs and preference for engine and winch.

Another page from the same booklet showing three designs of wooden trawler. These inshore boats have transom sterns to give a good working space aft. The middle boat, *Diana*, SY86, shows distinct Scandinavian characteristics.

Shemara, PD235, was one of the first fishing vessels designed by G. L. Watson after the resumption of the White Fish Authority 'grant and loan'. She was built by Jones of Buckie and was a consistent high earner for her owner, Jim Pitie.

FIFER
Built by Craftsmen with pleasure in mind

JAMES N. MILLER & SONS LTD.
St. Monance · Fife · Scotland

Phone: St.Monance (03337) 209   Grams: Miller, St.Monance

Although James Miller & Sons Ltd of St Monans were renowned for the quality of their fishing boats, in the 1960s they developed the Fifer range of motor sailers as an offshoot for the pleasure market based on their fishing boat experience. With a long, deep keel and cruiser stern, these boats came within the 28–35 foot range and were generally gaff rigged. Some were also built at other yards, such as James Noble in Fraserburgh. Many are still sailing.

Wooden boatbuilding alive in Macduff Shipyards Ltd in 2002. This 15 metre vessel was being constructed in the time-honoured way, using larch planking on oak frames and backbone. However, aluminium was being used for the wheelhouse and shelter deck.

The 21.3 metre *Adele*, BCK36, was built by Macduffs in 2001 and was about the maximum size they were able to build in wood. She was twin rigged for trawling for prawns. She was sold in 2007 and renamed as *Orion*, BF432, and continues fishing.

The transom sterned *Jasper II*, ex-PD174, had been built in Fraserburgh in 1981. When this photograph was taken, she was being sold and registered as BK7, working out of Eyemouth. She later was renamed at *Guiding Star*, FR897, before moving to Whitehaven as the *Renown*, registered in Lerwick as LK37. However, by then she was a very different looking boat with a full length shelter deck forward of the wheelhouse after the whaleback had been removed.

Decommissioning

Boats dumped higgledy-piggledy in Dundee's drydock in about 1995. These boats were, in the main, exceptionally seaworthy boats, built by skilled boatbuilders to last the test of time. Owners were paid a grant by the European Union to take their boat out of the fishing, the amount paid based on the size of the boat's VCU (Vessel Capacity Units), which is a mixture of engine horse power and boat size and reflects its fishing capability. Owners submit amounts that they are willing to accept and DEFRA decides who to accept and who to refuse. Boats then have to be chopped up, the keel being cut in half. Only then will they get their money, to spend on smaller yet more effective boats. To simply scrap them to appease faceless politicians seemed to most a crazy waste of a natural resource. The 40+ Fishing Boat Association was formed in 1995 to fight this government policy and they succeeded in saving a few boats but, in their own words, far too many of historically important vessels were scrapped. (*Photo: Rudiger Bahr*)

Two views of the fishing boat *Sara Jean*, PW58, built in Fraserburgh in 1955. By the 1990s she was working out of Newlyn and in 1997 was decommissioned alongside the scrapping quay at Newlyn, in the corner of the harbour. The work involved using a JCB to slowly hack its way through the internals of the boat and then attack the wooden hull. It was a painfully slow job, equally painful to watch, and a boat could take a week to dismantle and cart away to be burnt. Just an absolute waste. (*Photos: Billy Stevenson*)

Coiler and Capstan

Top: An 'Iron Man' on a Peterhead Zulu loaded with herring. Before the Iron Man was invented in Scotland, the drifters used a capstan that was turned by the crew using capstan bars slotted into the top. Strips of wood were nailed to the deck so that the crew, walking in circles around the capstan, kept a good grip on their feet. The Iron Man, which was more of a winch than a capstan, had a wheel which had to be turned by hand; nevertheless, it took some of the strain out of the hauling job. This later developed into the steam capstan.

Middle: The Millers Fifer 4-speed Many Purpose Winch on the *Pilot Us*, LK271 (see photo 3/9). It was said that this winch was the best available for fly-dragging seine-netting. Furthermore the manufacturer stated that it could also be used for ring-netting, drift-netting, purse-seining, lobstering, light trawling and line hauling. Various attachments aided the different fishing methods but this winch was regarded by many as their saviour, such was its capability. (*Photo: Mike Craine*)

Bottom: A Beccles steam capstan working away on the only remaining restored steam drifter/trawler, *Lydia Eva*, YH89, in 2011. This capstan was invented and produced by a partnership of William Elliott (mechanical engineer) and William Garrood (agricultural smith), based in Beccles, Suffolk. The first capstan was fitted to a fishing boat at Lowestoft in 1885 and within ten years they had sold 200. The steam capstan not only assisted in the hauling of the nets, which the fishermen were able to increase in length, but they also provided the extra power needed to sheet home the larger sails required to power the ever increasing size of the fifies and Zulus.

Fisher Folk

Whereas fishing methods and fishing boats are the tools of the fishermen, it is the social history of the fisherfolk themselves that in some ways is the true story of fishing. These fishers survived in their insular communities, strung out all along the east coast of Scotland and among the Northern Isles, and each had their own traditions and ways of living and fishing. Not until twentieth-century fishing displaced these communities in favour of large ports did these communities change over generations. Son followed father who had followed grandfather into fishing – there was generally no other choice. In far out regions, especially in Shetland, the fishers were beholden to their landlords, who forced them to fish – the choice was either to fish or be homeless.

Throughout much of Britain, fishing communities tended to be separate from the general populace, often set aside at the end of the town. In Scotland the villages tended to concentrate solely on fishing while the larger towns had fishing elements, although some had developed from what were initially fishing communities, especially on the Moray Firth. Even Edinburgh had its fishing areas, commanding much of the seafront – such as Port Seton, Cockenzie, Newhaven and Granton – though all the boats were registered under the Port of Leith, which was generally reserved for other shipping.

Communities weren't just about fishing. The boats had to be built, as we've already seen. Sailmakers and blacksmiths contributed, as did the various net-makers. Once the fish were landed, it had to be either sent straight to market or hawked about the locality or processed. Processing usually involved a either salt-cure or smoke-cure until the advent of refrigeration, deep-freezing and processes such as fish-finger production.

Boat maintenance was equally important. Here, John Budge and John Anderson are tarring the inside of an Orkney yole at Stromness in 1910. (*Photo: George Ellinson*)

Top: Other maintenance work included tanning the sails of the boats to prolong the life of the canvas, and barking the nets in a tannin solution to prevent the seawater rotting them. Tannin was obtained by soaking oak bark in hot water; later 'cutch', obtained from the acacia tree, was imported from the Indian subcontinent. Again, this was boiled up in big vats with water. Tanning solutions used the same materials, though often with dyes and 'special' additives to make the dyes hold such as dog faeces or rancid butter! In this photograph, taken in Lerwick, a bush rope (the rope between net and boat) is being treated with tar to stop their rotting.

Middle: Skipper Archie Smith sits in his small fifie yawl *Janet*, 412KY, in about 1880. The oilskins he is wearing were typical of the time, oiled canvas which was heavy and smelly. His leather boots would have carried him straight to the bottom if he fell overboard, which didn't really matter as hardly any fisherman could swim and lifejackets were almost never worn.

Bottom: Line fishing for cod and haddock was the oldest fishing upon which fishing communities had been established. The herring as a fishery didn't arrive until the end of the eighteenth century but soon became more profitable. In this photograph at Elie, Fife, (*c.* 1880) the hooks of the lines are being baited before the lines are redd into the wooden trays or baskets (i.e. laid in the trays/baskets in a tidy fashion to run out without snagging when setting). The bait was usually mussels, which had to be gathered before baiting could begin. Often the children would go down to the beach where mussels grew and pick them. These then had to be opened and the mussel itself removed. It was all a very tedious job and all the family would normally be involved.

Here at Auchmithie, the family are involved in pushing the boat out, although this photograph might seem a bit posed. However, it would often be that, once the boat was afloat, the wife of the fisherman would carry him out to the floating boat so that at least he wouldn't start out wet, even though the chance of him staying dry throughout the whole fishing trip was pretty slim. These two women certainly look pretty robust.

The crew of the steam drifter *Pursuit*, KY152, pose in 1910. Several are wearing the traditional fisherman's 'gansey' or Guernsey sweater. These were knitted by the fisherman's wife or mother with intricate patterns unique to each family. Thus, if a fisherman drowned and was later found, identification could often be made through the pattern of his sweater. Each fisherman is also wearing the characteristic flat cap, except the skipper on the right in the wheelhouse. Note again the above knee length leather boots.

The crew of the motorised drifter *Jeannie Mackay III*, WK197, of Helmsdale pose for the camera. The boat was built at the Thomson Yard of Buckie in 1949 and thus this photo was taken around the 1950s. Note how the clothing has altered, with rubber boots, fishing smocks and scarves replacing some of the awkward earlier clothing. The flat caps haven't changed though.

Fisherman J. Buttars getting a haircut aboard a boat moored on the East Pier at St Monans. Again, the fishermen are wearing ganseys, serge trousers and flat caps. There's a Beccles steam capstan aboard the boat (probably a fifie), which suggests the date to be 1890s/early 1900s. They all seem to be having a good laugh about something. (*Photo: William Easton*)

Above: By the late twentieth century, yellow oilskins had replaced the earlier type. Rubber gloves were also worn to offer some protection to the hands. Here a trawl net has been emptied and the men will begin to sort and gut the fish.

Left: Side trawling, where heavy bobbin trawls were to a large extent manually heaved aboard a vessel, had almost disappeared from Scottish waters by the beginning of the 1990s. Power blocks and hydraulic net haulers replaced human muscle but there were still plenty of dangers in fishing. Beam trawling was introduced to the Scottish fleets by a Welshman who moved up to Buckie and quickly took hold in the late 1980s. (*Photo: Peter Brady*)

Discharging herring from the hold of a boat at Wick using a quarter cran basket. These baskets were made by local basket-makers to a set size (17¼ inch top diameter, 14¼ inch bottom diameter and 14¼ inches in height) and set design, the finished basket being checked and having a branded piece of hardwood beneath each handle. The cran had been set in Scotland in 1889 as 37½ imperial gallons, so that a quarter cran was approximately 180 fish or 3½ cwt of fish. This photograph just shows the amount of herring that was landed.

Stream drifters jostling for space at the quayside in Wick. In the foreground a quarter-cran basket of herring is being swung ashore. The steamer *Valor Crown*, BF249, was a common visitor to Wick. The capstan is set well forward and the mass of net shows just much each boat set – at least two miles worth. It was said that if all the nets set out in the North Sea were put together, they would reach across the Atlantic Ocean.

The fish market at Pittenweem. Fish is laid out on the quayside ready to be sold. Today, as in all landing ports, modern market halls with electronic auctioning are the reality.

Aberdeen, 1914, with fish ready to go to the curing yards. Wooden fish boxes were used for many years until the late twentieth century, when plastic boxes became the accepted container to store fish in. These boxes contain herring and fish curer Andrew Masson is loading up his vehicle. The curers exerted huge influence upon the industry, often dictating the price to the fishermen and employing bands of women to gut the fish.

Herring gutting at J. & M. Shearers' herring curing yard in Lerwick in the mid-twentieth century. The herring lassies were Scottish women or girls employed by the curers to gut the herring. They followed the fishing fleets down the coast so that they would work in Shetland in the spring and end up in Lowestoft in the autumn. Some would go to the west coast and work from Stornoway down to Mallaig. They were regarded as one of the toughest groups of working women, gutting from the early morning to the evening, six days a week. The fish were loaded into long troughs called 'farlanes', gutted by a quick movement of the knife and then laid into barrels with layers of salt. They were generally paid by the barrel of fish prepared – two gutters and one packer worked as a team – although they did receive a small amount of money prior to their being taken on.

The quay at Shearer's yard is full of barrels. Coopers were kept busy almost constantly, making the barrels from timber imported from Scandinavia. Once filled, they were left to settle before being topped up. Fishery Officers would inspect the barrels when opened and approved lots would be branded as a sign of 'Scotch Cure' quality. The barrels were exported all over the world, especially the Eastern European market. In the early eighteenth century it was exported to Africa and the sugar plantations in the Caribbean to feed the slave workers. The Shetland fifie *Swan*, LK243, can be seen in the background.

FILLING HERRING BARRELS WITH BRINE

Above: A group of herring lassies in the 1920s. These remarkable women slept in hostels which were arranged by the curers. Often they were dormitories where twenty or thirty could sleep. They had Sundays off and many were church-goers. Walks about in the afternoon often heralded relationships with local fishermen. The women were also renowned for their singing while at the farlanes.

Left: A poignant photograph of Gracie Stewart topping up the brine in a barrel. Although dressed in clothing that must surely stink of fish, and doing one of the muckiest of jobs, this picture depicts her as a soft and gentle person. The author met her grand-daughter by chance in Portsoy in about 1998 and she furnished him with the following information. Gracie was 18 when this photograph was taken and working in Lowestoft, where she later met her husband and had two children. Her pay was 1/2d an hour with another 10s at the end of the season.

Coopers at work. Making herring barrels was one of the largest industries associated with fishing and thousands of coopers were employed. A good cooper could make seventy barrels a week. Most were made of Scandinavian spruce, though earlier local larch and birch was used; the wood had to be dry and heavy. The staves of the barrel had to be cut and planed to exactly half an inch thick. If the joints were not straight, they would not fit closely together and the barrel might leak. The staves were held together by wooden hoops and were moistened inside with water and heated until the cooper judged them the right temperature to bend into shape. An adze was then used to cut the rim or 'chime' of the barrel. Once the rim had been cut, a special tool called a croze was used to make the groove in the lid. The 'head' or lid was then fitted on using a chive and flencher.

Hundreds of barrels in a curing yard. Strict regulations covered the making of barrels and their use in the fishing industry. In 1815 it was ruled that barrels staves had to be exactly half an inch thick, and the exact number of hoops to be used was also set down. If the barrels did not meet the requirement, pickle could leak out or air could get in, which would contaminate the fish and spoil the cure. To stop the use of undersized barrels, a standard capacity of 32½ imperial gallons for a barrel was introduced in 1889. A full barrel held a cran of herring (a volume measure of around 1,000 fish). The cooper worked closely with the curer to ensure that his barrels met with these regulations and were watertight.

Left: Women gutting and splitting herring for kippering. Normally the best herring was sent to be smoked. Various different smokings produced a variety of products – from the kipper to the red herring. Golden herrings, which were (and still are) heavily salted, were sent to the Mediterranean countries. Red herrings were ones left in the top of the smokehouse for weeks so that they smoked and then sweated between smokes. They really were strong in flavour and the expression 'red herring' comes from the fact that they were known to put a hound off its scent.

Below: The Finnan haddie is cold smoked haddock and originated in the north-east of Scotland. Some say this was in Findhorn on the Moray Firth, while others insist it was Findon, near Aberdeen, where it has been a popular dish since at least 1640 and probably a lot longer. Londoners fashioned it in the 1830s but, given it only received a light smoking, it only lasted three days at most.

Above: The Arbroath Smokie, on the other hand, is hot smoked and its birthplace is clear: the small village of Auchmithie, three miles north of Arbroath. Various theories have been put forward as to its origins, including one about a house where haddock had been left to dry burning down! In this photograph, a local fishwife is smoking the fish on sticks laid over halved whisky barrels with fires underneath. Layers of coarse sacking of locally produced jute would then trap the smoke in. In the nineteenth century many of the villagers moved to Arbroath and took their delicacy with them. Thus today's 'Arbroath Smokie' follows a process which is typical of similar smoking processes carried out to this day in Scandinavia, where it is said Auchmithie's original inhabitants had themselves originated.

Right: Fresh fish was hawked around the locality in which it was landed by the fishwives, this one being in St Andrews. Sometimes these women would walk 20 miles in a day with heavy loads of fish on their backs.

Along the Coast

From tiny beach settlements where boats were dragged ashore to escape the surge of the high water to sheltered stone-built harbours capable of holding a thousand small boats, the east coast of Scotland has it all. Of course, prior to the eighteenth century Scotland suffered a distinct lack of harbours although, with fishing curing establishments being opened at various places along the coast – such as Wick and Fraserburgh in 1810, Helmsdale in 1813 and Peterhead in 1820 – fishing ports were established. However, small beach-based fleets continued throughout the nineteenth century, and even today in the twentieth-first, a few small creel and coble fishers still prefer to work off beach havens such as Catterline. When, between 1880 and 1882, the government allocated £7 million for harbour building, only a tiny proportion was spent in Scotland – £500,000, much of which went to improving the harbour at Anstruther.

Berwick-Upon-Tweed to the Firth of Forth
St Abbs harbour – once called Coldingham Shore – with several fifies going out. Small harbours afforded protection to the growing fishing fleets in the nineteenth century. Before that, they were at the mercy of the sea both during the fishing and at a time when they might want to head home to shelter from an approaching gale. Even so, many boats were often lost when approaching the harbour in the eye of a storm or at night. Fishing has never been a safe occupation and continues today to be a dangerous job. (*Photo: Campbell McCutcheon*)

St Abbs had its harbour built in 1832, with an adjoining harbour built in 1890. This was extended probably due to the Eyemouth fishing disaster of 1881, when 129 fishermen and twenty-three boats were lost along this coast during a particular ferocious storm. Most were from nearby Eyemouth but the tiny fishing communities of Burnmouth and Cove suffered the loss of thirty-five while St Abbs lost only three men. Newhaven and Fisherrow lost twenty-two men. This remains one of the worst fishing losses in British history.

Dunbar harbour by William Daniell. Daniell travelled right around Britain between 1814 and 1825 and produced a host of wonderful aquatints of harbours, boats and seascapes. Dunbar was once one of the most important harbours in Scotland, being built about 1650, and has a long fishing tradition. Dutch 'busses' unloaded their catch there at a time when the town bustled with shipyards, rope works, sailmakers and foundries. The old harbour is called the 'Cromwell' harbour as it was built upon his instructions, while the new harbour is known as the 'Victoria' harbour as she was on the throne.

A fifie leaves Port Seton harbour. Both Port Seton (built *c.* 1880) and nearby Cockenzie (built 1830) were home to a substantial herring fleet. Cockenzie was built as a coal harbour for the East Lothian collieries. Other fishing communities along this part of the coast were at Prestonpans, Musselburgh and Fisherrow. Prestonpans's harbour at Morrison's Haven has gone while Musselburgh's harbour (sometimes referred to as Fisherrow's harbour – it's hard to know which is right as the two communities are almost on top of each other) is very small.

Newhaven was the home of the Edinburgh fishermen. The first settlers were Flemish and, in 1572, a Society of Free Fishermen was set up on the lines of a Flemish guild. The Newhaven fishwives were famous throughout Scotland and even George VI was said to have remarked that they were the handsomest women he'd ever seen. When a group travelled to London for the international fisheries exhibition in 1881, their costume was so impressive that it soon became the fashion of the day. They continued hawking fresh fish around Edinburgh up to the 1950s.

Top: Newhaven today, with the small fifie *Robina Inglis*. It's a drying harbour and is dwarfed by the Port of Leith on one side and Granton harbour a mile west. Granton was built as a model port and ferry terminus and was opened on the day of Queen Victoria's Coronation in 1838. However, it never seemed to achieve its aims, though it did become an important fishing port and rivalled Aberdeen in the early part of the twentieth century. In 1928 there were sixty-two steam drifters working from there.

East Neuk of Fife to Arbroath

Middle: A motorised fifie lying in St Monans harbour in the East Neuk of Fife. The East Neuk consists of a string of fishing villages – Elie, St Monans, Pittenweem, Anstruther, Cellardyke and Crail – though only Pittenweem has survived as a fishing harbour where today mostly prawns are landed. St Monans was home to the renowned boatbuilding yard of James Miller & Sons, who built hundreds of fishing boats between 1747 and the yard's closure in the 1990s. (*Photo: Scottish Ethnological Archive*)

Bottom: Two boats at Pittenweem. Left is *Progress*, KY143, while on the right is *Courageous II*, KY59, built as a ring-netter by Weatherheads in 1934. Today she has been thoroughly restored and is kept in Bristol harbour, from where she is sailed all over the west coast of Britain, Ireland and over to France.

Above: The small clinker-built fifie yawl *Rose Leaf*, KY160, sits atop the quay in Pittenweem in 2001. Fitted with an engine, she fished with lines and creels for many years before ending up being used to fish a few creels as a pleasure pastime. Still, she is a lovely example of these boats and can often be seen out of the water in the off season. Pittenweem's modern fishmarket lies hidden by the boat. (*Photo: Mike Craine*)

Left: Cellardyke harbour is today overshadowed by the nearby and much larger Anstruther, although it is much older. Indeed, it is easy to miss it altogether! Until the harbour at Anstruther was built in the nineteenth century, Cellardyke was the centre of the Fife fishery for many a generation. Today Anstruther is home to the Scottish Fisheries Museum, arguably the best fishing museum in Britain.

The harbour of Crail lies at the seaward end of the East Neuk of Fife and is ancient. It is said that fish was exported from here to mainland Europe in the ninth century and that the Dutch learnt the art of salting fish here as well. The harbour entrance is notoriously difficult to negotiate and can be closed off with baulks of timber in heavy weather. Here, boats are gathered by the harbour entrance and three fifies can be seen drawn up. Small boats are being off-loaded by horse and cart.

Today, picture-postcard Crail is home to creel fishers such as the Reilly brothers, though they land much of their catch at Pittenweem. However, cooked crabs and lobsters can be bought at the small shack in season.

A crowd of on-lookers gaze down at the Dundee-registered small fifie – or bauldie – 520ME at St Andrews, though it is impossible to see what the crowd is waiting for. Most of the men appear to be fishermen, judging by their attire. On the harbour wall are some lobster creels which appear to be framed in metal while there are also the iron shoes and cross beam of a beam trawl. St Andrews was a harbour before the Reformation and conducted trade throughout northern Europe. Every April the Easter Senzie Fair was held and what is thought to be the remains of market booths can be seen in the old walls by the harbour below the Priory.

The docks at St Andrews with a host of fishing boats lying idle. Timber is being unloaded from the schooner, quite likely having come from Scandinavia. All the small fishing boats are registered in Dundee. Timber lying on the foreground could well be boatbuilding timber, though there is no mention of a builder working in this area. There are sluices gates between this inner harbour and the outer harbour which can be closed at times, though their main use seems to be to clear sediment from the outer harbour by flushing out.

Arbroath to Peterhead

Two boats on the slipway at Arbroath. On the left is *Ajax*, AH96, and behind is *Fortuna*, AH153. In the late eighteenth century, few fishermen worked from Arbroath even though there was a harbour. It wasn't until the 1830s that the Auchmithie fishers were enticed away from their village to the town with the promise of land to build on. When they did come, they settled at the end of the High Street, in what is known still today as the 'Fit o' the Toon'. It is around here that the Arbroath Smokies are mostly made. Today, the last white fish trawler, John Swankie's *Crystal Tide*, has given up fishing in 2013, though there remains a small fleet of inshore boats in Arbroath and Mackays Boatbuilders remain working from the slipway. (*Photo: Edward Valentine*)

Here, various colourful fishing boats are moored in Gourdon harbour in the 1950s. The harbour was designed and built in 1820 by Thomas Telford and extended during the mid part of the century. 8,000 barrels of herring were exported from here in 1881, though this fishery declined in the early 1900s. Thus the fishermen were among the first to adopt diesel engined boats for long-lining, abandoning their steam drifters, and supposedly they were the first fishermen in Scotland to commercially work long-lines.

Four boats in Stonehaven harbour in the 1960s. From left to right, they are: *Sapphire*, A741; *Sweet Promise*; unloading inside is *Scotch Lass*, A72; and outside *Grateful*, A753; and forward is *Superb*, A611. Stonehaven has had a harbour since the early sixteenth century and the town was later laid out as a fishing station, though it has never really enjoyed success as one, even though it was the only safe harbour between Arbroath and Aberdeen. In the nineteenth century there were only a handful of boats working from the harbour. (*Photo: Edward Valentine*)

Another view of Stonehaven harbour in the mid-1960s, with *Grateful* outside of *Sapphire* and the *Mary Gowans* on her own. Stonehaven only enjoyed a boom during the end of the nineteenth century and into the twentieth, during the herring boom of that time. However, afterwards it was only home to a small fleet and part of the trouble was said to be the silting up. (*Photo: Edward Valentine*)

A postcard view of Aberdeen with two steam trawlers passing each other. Aberdeen's fishing fortunes didn't really commence until a syndicate of locals bought a steam tugboat, the *Toiler*, in 1882 to tow a fish trawl, to much opposition. However, as we shall see in various instances over this series of books, fishermen were quick to oppose but equally quick to join in if they saw good results. By 1900 there were 205 steam trawlers and thirty years later this had increased to 270 trawlers and forty steam liners. Although boats do still land there, Aberdeen's fishing has been dwarfed by oil.

Cleaning haddock at Aberdeen. Again, it was more often than not women who gutted and cleaned whitefish. Much of the haddock would be sold fresh, though smoked haddock was always a favoured dish. The fellow with the black cap seems to be intent in keeping an eye out upon the women workers. (*Photo: Campbell McCutcheon*)

Above: Some twelve miles north of Aberdeen is the harbour of Collieston. This photograph, taken about 1890, shows various small open fifie type boats drawn up on the beach. Fishermen's small cottages cluster around the beach in an irregular pattern while some larger, two-storied houses suggest ownership by wealthier fishermen who probably also owned the boats. Fishermen's sheds can also be seen, where some of the haddock would have been smoked.

Left: On a promontory a bit north of Collieston is the Old Castle of Slains, a corner section of the castle after it was blown up by James V in the early seventeenth century after a Catholic uprising. Below, working from a small sandy beach, some fifty fishermen earned a living, some living in these up-turned boats, although more substantial stone buildings had also been built.

William Daniell's view of Peterhead. In 1822 he mentions that a new harbour had recently been built, this being the North Harbour that was completed that year, designed by Telford. But the 'old harbour was originally a shallow cavity in the rock'. The new harbour consisted of 'about thirty-two acres protected by a spacious pier of about five hundred from the land, with a returning head of one hundred feet; from the base of the pier extends a new wharf wall, nearly one thousand feet; and at a right angle to it is an extensive dry-dock for repairing of ships, cut out of the solid rock'.

Fishmarket in Peterhead. Although some of the fishers in Daniell's time were following the herring, it wasn't until the year after he'd visited, in 1823, that the herring gained over whaling. And again its growth was rapid: by 1836 there were 262 boats there and 40,000 barrels of herring were being exported. By 1850 it was second only to Wick in the north, having twenty-seven curing facilities in the town. Here, in the corner of the south harbour, was a small beach where some of the boats would unload their catch and sell. The big warehouse, on the corner of Harbour Street and Union Street, still survives.

The south harbour. This photograph is believed to date from the 1920s, when the harbour had been greatly extended. Today, Peterhead is the largest white fish and pelagic fish harbour in Britain, even after several rounds of decommissioning has depleted its fleet. Today's fishmarket is 1,250 feet long and runs along the East Pier and the Alexandra Basin. However, oil also now plays a major part in the town's economy, with a large quay dedicated to serving large ships.

Fraserburgh to Inverness

Fraserburgh harbour developed from Alexander Fraser's sixteenth-century settlement into one of Scotland's major fishing ports. Originally the fishers worked from the tiny village of Broadsea, just west of Kinnaird Head, but began to use Fraserburgh's beach in the seventeenth century. The first harbour dates from 1832 and various projects increased it in size over a century. Curing of fishing started in 1810 and nowadays it boasts a fleet of big pelagic trawlers. The fish market was upgraded in 2007 to included fully refrigerated facilities, signifying the start of a major expansion of fishing at Fraserburgh in pelagic, demersal and nepthrop species.

A Nairn-based Zulu among others at Fraserburgh in about 1880. Nets are being repaired and the general clutter about the deck of this boat gives a picture of just how hard the work was. The hold appears full of netting. Ropes are scattered about the deck. By the mizzenmast is the crutch into which the mainmast was lowered when the boats were lying to their nets.

The tiny village of Pennan was a prime example of a small fishing community, described by Peter Anson as having the most picturesque position on the whole of the east coast. It's a village sandwiched in between the sea and the cliff and consists of a row of houses lining the road. It was the location used for the hotel and village in David Puttnam's film *Local Hero*.

Several other small villages cling to the cliff, such as Crovie and Gardenstown, both of which once had their own fishing fleets. Macduff has one of the best harbours along the coast and has had associations with the herring fishing since the early nineteenth century. In 1921, it was one of the first to adopt the seine net. Today it is home to Macduff Shipyards Ltd.

Banff lies a mile across the bay and rivalled Macduff. It had a healthy herring fleet until the harbour silted up in the middle of the nineteenth century and many of the boats went to Macduff.

Whitehills lies around Knock Head and consisted of a quay dug out of the rock which protected its sizeable fleet which, in 1880, was 158 strong. However, the present harbour was built in 1900 and a considerable fleet continued to use it in the twentieth century. It is said that some of the inhabitants can trace their families back over 300 years in the village.

The small harbour of Portknockie was home to some of the enterprising fishermen in the Moray Firth who, it seems, settled there from Cullen in the late seventeenth century. During the herring boom it was a busy little place and the harbour is said to have been the deepest north of Aberdeen.

Top: Buckie harbour in the 1950s. Buckie was originally the hamlets of Ianstown, Yardie and Buckpool, the inhabitants of which fished great lines away from the coast. The first harbour was made from timber in 1843 and, after this was swept away by a storm, another was built ten years later. In 1877 the present large harbour was built and there were over 500 Zulus based there in 1900.

Middle: Buckie was also home to various boatbuilding firms, the most renowned being Herd & Mackenzie, Jones of Buckie, Thomson's Yard and the McIntosh Yard. Today, Buckie still has its fishing fleet and some industry associated with fishing. It is also home to the Buckie & District Fishing Heritage Centre, which has a splendid historical display and archive on local fishing.

Bottom: Nearby Portgordon's small harbour was once packed with fishing boats – here fifies and Zulus together – even though it was initially built to export grain from the local farms. At one time it was busier than Buckie, but in time the fishing boats moved there.

Portgordon lies at one end of Spey Bay. At the mouth of the River Spey a salmon fishery has been carried on for generations and the Tugnet ice house, built in 1830, still survives. It is the largest ice house in Scotland and has six vaulted chambers with cobbled floors and drainage sumps. The ice was collected from the river during the winter and would last several months. It is said that 150 men worked here, collecting ice, fishing in season and loading ships with ice and fish before it was exported to London.

LOSSIEMOUTH HARBOUR.   202,722 JV.

Lossiemouth lies at the other end of Spey Bay. Its harbour dates from 1830, at a time when some forty-five boats were based here. Steam drifters landed here. Lossiemouth's one claim to fame is that the first Zulu, *Nonesuch*, was built here by William Campbell in 1879, combining the best of both scaffie and fifie. By the 1880s, it has been said that there were 3,665 Zulus registered in Scotland.

Lossiemouth was also the first port to adopt the seine net. John Campbell also designed and launched the first modern seine-net boat.

William Daniell's 1821 aquatint of Burghead depicts a substantial harbour. He states that the pier was built eleven years previous, i.e. in 1810. In his time it was already a herring station of note and several curing yards surrounded the harbour. However, it seems that the harbour was never fully used for, in 1840, there were only forty-three boats based there.

The Black Isle to Duncansby Head

A scaffie leaves the outer harbour of Avoch. Avoch fishermen are not of Highland origin – there is disagreement as to whether they have Spanish, Welsh, Cornish or Norse roots. In winter they fished the upper Firth small herring, which were renowned for their sweet taste. In April they would sail up the Caithness coast to long-line for whitefish, and in summer they would fish the herring in either Caithness or on the west coast, after which they might catch lobsters and crabs. These full-time fishers followed fish throughout the year. Avoch fishermen were also known for their wedding ceremonies, which took place on the Friday so that they could spend the whole weekend in celebration.

The River Brora was one of the finest salmon fishing rivers in Scotland and the town grew up around its mouth. Fishing was once the main occupation, though by the middle of the nineteenth century they were only fishing the seasonal herring. Most of their time was spent in the surrounding fertile fields. Today Brora, like its neighbour Golspie, is a popular holiday resort. Fishing now seems only recreational.

Helmsdale too grew up around the mouth of that river and the original settlers were displaced crofters from the Sutherland Clearances. The Duke of Sutherland provided fishing facilities – the harbour, village and curing yards – and the herring boom brought over 200 boats to the harbour in season. Here is Daniell's aquatint of 1821 and he states that the fishery started here in 1821.

Helmsdale might have been a busy fishing station when this photograph was taken in the 1880s, with several big boats unloading fish, barrels lining the quay and a couple of schooners waiting to export cured herring, but it had great difficulty competing with the other bigger ports, such as Wick to the north and the Moray Firth southwards. Even so, it did have a small fleet of seine-net boats in the 1930s and fishing does continue today, albeit on a small scale.

Surprisingly, Lybster became the third largest herring station after Wick and Fraserburgh in 1840. However, fifty years later the herring boom was almost over and when steam drifters arrived, they tended to work out of Wick. In 1877, a government Commission found that 'Lybster Harbour is in constant danger of being totally destroyed and with it the means if subsistence of nearly the whole population of the place, who are almost entirely dependant upon the fisheries'. This was after damage to the storm-lashed harbour; although repairs were made, it wasn't until 1950 that the inner harbour was built and the outer one in 1984. It remains a working port, though, with plenty of recreational boats alongside the few that work.

At Hempriggs, Daniell found 'herring-boats drawn up, and the people employed in barrelling fish. The continual screaming of gulls and cormorants caused a harsh dissonance, which (if the Iricism may be pardoned), accorded well with the wild scenery. This remains a wild coast today and one might be forgiven in wondering just how the fishers used these coves for fishing. At Whaligoe, a few miles south of Hempriggs, the harbour (if you can call it that) is connected to the village by 365 steps down the Cliffside. Seeing it today it is amazing to think that there were 35 boats here in 1850, a curing station and that women carried the catch up the steps and the six miles to Wick. It seems that, to secure their boats from gales, they 'hang up their yawls by ropes on hooks fixed in the surface of the rock above the level of the water, where they are safely suspended till the weather is fit for going to sea', according to the Old Statistical Account.

Wick was the herring capital of Europe in 1865, with over a thousand boats landing. Some 3,800 fishermen were employed, with another 4,000 curers, coopers, gutting girls, carters and labourers on shore. Some 500 gallons of whisky were consumed in a day when the fishing was good. The herring girls must have added colour to what was otherwise a tedious and arduous job. (*Photo: Campbell McCutcheon*)

But in the early twentieth century, the fishing fortunes declined and although fishing is still an occupation in Wick, large fishing boats are seldom seen and those that do fish set creels more than nets and lines. These seine boats too are a thing of the past.

In 1855 Keiss was home to forty-nine boats, manned by 180 fishers, keeping ninety-six gutting girls, seven coopers and 190 netmakers busy, even though the village is only a few miles away. It later became an important crab landing harbour, one of only a handful on the east coast and the nearest being at Crail. The remains of the curing station and ice house can still be seen.

The Northern Isles

Boats in Stromness, Orkney. The natural harbour here made the place popular with passing whale boats making for the North Atlantic between 1790 and 1830. After that, it became a fishing station with curing yards associated with the herring fishery. Orkney is home to Skara Brae, a preserved prehistoric fishing village occupied between 3100 BC and 2500 BC.

The isle of Stronsay owes much to the herring fishery; in the nineteenth century, people were encouraged in when Samuel Laing set up his herring station in 1816 with six fishing boats. Whitehills soon became the main town and 300 boats landed here and 12,000 tons of herring were landed here. (*Photo: Campbell McCutcheon*)

A herring fishing station at Holmsgarth, Lerwick. Much of Lerwick's herring was sent to Germany. Lerwick had thirty-five curing stations in 1901 and there were 141 in Shetland as a whole. Several boats from the Isle of Man are anchored off.

Large first-class boats lined up at Lerwick. Even at the end of a hard day, the men found time to get 'A Day Ashore'!